THE POETRY OF CADMIUM

The Poetry of Cadmium

Walter the Educator

SKB

Silent King Books a WhichHead Imprint

Copyright © 2023 by Walter the Educator

All rights reserved. No part of this book may be reproduced in any manner whatsoever without written permission except in the case of brief quotations embodied in critical articles and reviews.

First Printing, 2023

Disclaimer
This book is a literary work; poems are not about specific persons, locations, situations, and/or circumstances unless mentioned in a historical context. This book is for entertainment and informational purposes only. The author and publisher offer this information without warranties expressed or implied. No matter the grounds, neither the author nor the publisher will be accountable for any losses, injuries, or other damages caused by the reader's use of this book. The use of this book acknowledges an understanding and acceptance of this disclaimer.

"Earning a degree in chemistry changed my life!"
- Walter the Educator

dedicated to all the chemistry lovers, like myself, across the world

CONTENTS

Dedication v

Why I Created This Book? 1

One - Cadmium, A Paradox 2

Two - Until The Very End 4

Three - Double-edged 6

Four - We Must Take 8

Five - Lesson In Light 10

Six - Crystal Clear 12

Seven - Forever Tame 14

Eight - Unforeseen 16

Nine - Cadmium's Power 18

Ten - Caution And Care 20

Eleven - Masterpiece In Bloom 22

Twelve - Cadmium's Grace 24

Thirteen - Held Light 26

Fourteen - Endless Possibility 28

Fifteen - Same Roof 30

Sixteen - Venomous Prize 32

Seventeen - Think Twice 34

Eighteen - Lesson To Convey 36

Nineteen - Cautionary Flair 38

Twenty - Captures The Eyes 40

Twenty-One - Balancing Your Gifts 42

Twenty-Two - Delicate Gleam 44

Twenty-Three - High And Low 46

Twenty-Four - Cadmium, Oh Cadmium . . . 48

Twenty-Five - Friend And Foe 50

Twenty-Six - Respecting Its Power 52

Twenty-Seven - Secret To Keep 54

Twenty-Eight - Deepest Awe 56

Twenty-Nine - Wisdom Control 58

Thirty - Day And Night 60

Thirty-One - Toxic Trail 62

Thirty-Two - Gold And Silver 64

Thirty-Three - Intricate Schemes	. . . .	66
Thirty-Four - Listen And Rejoice	. . . .	68
Thirty-Five - Cadmium, The Enigma	. . . . .	70
About The Author	. . . .	72

WHY I CREATED THIS BOOK?

A poetry book about Cadmium can serve as a vehicle for artistic expression while also fostering a deeper understanding and appreciation of the chemical element, its properties, and its role in the world.

ONE

CADMIUM, A PARADOX

In the depths of the Earth, I lie concealed,
A metal rare, with secrets unrevealed.
Cadmium, they call me, a silent force,
With a lustrous sheen, a captivating source.
 Hidden in ores, my beauty does reside,
A soft, bluish-white, where wonders hide.
With atomic number forty-eight, I stand,
A symbol of transformation, in demand.
 From batteries to pigments, I find my way,
In vibrant hues, I bring life to display.
Yellow, orange, and red, I paint the sky,
In art and industry, my colors never lie.
 But beware, for I hold a toxic touch,
A warning sign, a danger that's much.

My presence in the air, the water, the soil,
A silent poison, a life I can foil.
    Yet, amidst the darkness, hope persists,
For scientists strive to find the twists.
In solar cells and biomedicine's quest,
My properties explored, for the best.
    Cadmium, a paradox, both friend and foe,
A symbol of progress, a tale to sow.
In the realm of elements, I proudly stand,
A reminder of nature's intricate hand.

TWO

UNTIL THE VERY END

In the realm of elements, a paradox is found,
A metal with secrets, both silent and profound.
Cadmium, its name, whispers a tale,
Of beauty and danger, an uncertain trail.

With a shimmering hue, like the sun's golden weave,
Cadmium adorns, a sight to conceive.
In pigments it dances, creating art's flare,
A master of colors, beyond compare.

Yet beneath its allure, a darkness resides,
For Cadmium's touch, beware as it hides.
In batteries it thrives, with energy untamed,
Powering devices, where life is ordained.

But caution, dear souls, for Cadmium's embrace,
Can bring forth a curse, a toxic disgrace.
The "ouch-ouch disease," it whispers with dread,
A reminder of dangers, where health fears tread.

Yet even in peril, Cadmium finds a way,
In solar cells it shines, harnessing the day.
And in biomedicine's realm, it aids the fight,
A paradoxical element, both shadow and light.

   Cadmium, a reminder of life's complexities,
A friend and a foe, in equal capacities.
For within its existence, we witness the blend,
Of beauty and danger, until the very end.

THREE

DOUBLE-EDGED

In Cadmium's embrace, a tale unfolds,
A shimmering beauty that nature beholds.
A lustrous metal, a radiant hue,
Cadmium, a treasure, both old and new.
    Born in the depths of Earth's ancient core,
A gift from the stars, forevermore.
Its brilliance adorns the artist's brush,
Creating pigments with a vibrant rush.
    But beware, for Cadmium holds a secret,
A toxic touch, a danger, if we permit.
In batteries it powers our modern age,
Yet its toxicity, we must engage.
    A paradox it seems, this element rare,
Both beauty and danger, a delicate affair.
In solar cells, it seeks to shine,
Harnessing sunlight, a power divine.

In biomedicine, it shows its worth,
Fighting diseases, bringing healing forth.
But handle with care, this double-edged sword,
For Cadmium's allure should not be ignored.

A paradox it remains, this element of might,
Its glow and poison, a dance in the night.
In science's realm, it plays a role,
A reminder of the wonders that we behold.

So let us tread lightly, with reverence and grace,
In Cadmium's presence, a delicate embrace.
For in this paradox, we find a truth,
Beauty and danger intertwined, in our youth.

FOUR

WE MUST TAKE

In the realm of the periodic table's dance,
A beauty hides, in Cadmium's expanse.
A metal rare, with lustrous grace,
Its paradoxes leave us in a daze.
   Oh, Cadmium! A friend, a foe,
With toxic touch, a warning we must know.
In batteries, you power our lives,
But handle with caution, for danger thrives.
   Pigments of yellow, red, and orange,
You grace the canvas, an artist's range.
Solar cells harness your radiant might,
But tread with care, for toxicity's bite.
   In biomedicine, you lend a hand,
Aiding diagnostics, a cure unplanned.

But in your essence, a cautionary tale,
A delicate balance, where dangers prevail.
    Cadmium, a paradox, a mystery profound,
In brilliance and toxicity, you astound.
Handle with reverence, grace in your wake,
For beauty and danger, a bond we must take.

FIVE

LESSON IN LIGHT

In the depths of the Earth, a hidden treasure lies,
A metal of paradox, Cadmium, with shimmering guise.
A friend and a foe, this element so rare,
Its beauty and danger, an intricate affair.

In batteries it dances, a mighty force it wields,
Powering our devices, the energy it yields.
Yet caution we must exercise, for toxicity it bears,
A delicate balance, a warning it declares.

In pigments it thrives, with colors so bold,
Painting the world, a sight to behold.
But tread with care, for the dust it may release,
A poison to the lungs, a danger that won't cease.

In solar cells it gleams, harnessing the sun's light,
A symbol of progress, a sustainable delight.
But handle with reverence, for its touch may bring

harm,
A reminder of nature's delicate charm.
　In biomedicine it offers hope, a cure it may bring,
Fighting diseases, a healing it will sing.
Yet handle with grace, with caution and care,
For Cadmium's power, a double-edged affair.
　Oh Cadmium, a paradox in your core,
A friend and a foe, forevermore.
In your beauty lies danger, in your power lies might,
A reminder of balance, a lesson in light.

SIX

CRYSTAL CLEAR

In the depths of the Earth, where secrets lie,
There dwells a metal, both friend and foe,
Cadmium, they call it, with its shimmering glow.
   A paradox it is, this element of grace,
With hues of yellow, orange, and red,
It dances with beauty, yet hides its dread.
   In art, it finds its purpose and might,
A master's brushstroke, a canvas alight,
Cadmium, the muse, guiding the artist's flight.
   But tread with caution, for danger awaits,
For Cadmium's touch can seal one's fate,
In industry's grasp, it bears its weight.
   Oh, Cadmium, you're a double-edged sword,
A blessing and curse, both revered and abhorred,
A reminder of life's complexities, adored.

In solar cells, you harness the sun's might,
Unleashing energy, a source so bright,
Cadmium, the conductor, guiding us towards light.

In biomedicine, you lend a helping hand,
Fighting disease with your healing band,
Cadmium, the warrior, protecting life's grand.

Yet, we must handle you with utmost care,
For your toxic nature, we must beware,
Cadmium, the enigma, a delicate affair.

So, let us honor you, Cadmium dear,
With reverence and grace, we hold you near,
A reminder of life's fragility, crystal clear.

SEVEN

FOREVER TAME

In depths of Earth, a secret lies,
A metal bold, that both beautifies,
And harbors danger, in its core,
Cadmium, paradox galore.

A shimmering hue, like golden thread,
It paints the canvas, where colors spread,
From vibrant yellows, to fiery reds,
Cadmium's touch, a masterpiece it sheds.

In solar cells, it finds its place,
Harnessing light, with gentle grace,
Powering homes, with sun's embrace,
Cadmium's energy, a saving grace.

In biomedicine, a silent knight,
Fighting diseases, with all its might,
Cadmium's whispers, bring healing light,
Aiding us in the battle, day and night.

But heed the warning, in its gaze,
For Cadmium's beauty, a dangerous maze,
Toxicity lurking, in its embrace,
Handle with caution, embrace its grace.

A dance of balance, this element's game,
A reminder of life's intricate frame,
Cadmium, a paradox, a flame,
Beauty and danger, forever tame.

EIGHT

UNFORESEEN

In the depths of the Earth, a metal dwells,
Cadmium, the enigmatic tale it tells.
A shimmering beauty hidden from the eye,
A paradox of grace and danger, oh so sly.
　In solar cells, it dances with the sun,
Capturing light, a power yet to be outdone.
Harnessing energy, a celestial embrace,
Cadmium shines with elegance and grace.
　But beware, dear mortal, its toxic might,
A venomous touch that can cause endless blight.
In biomedicine, it walks a fine line,
A double-edged sword, a delicate design.
　Its hues of yellow, like a golden ray,
Illuminate the world in a vibrant display.

Yet in its depths, a darkness does reside,
A silent reminder of the dangers it hides.
   Cadmium, a paradox, a puzzle untold,
A testament to nature's secrets, bold.
Handle it with care, with reverence and awe,
For it mirrors life's complexity, raw.
   So let us marvel at its dual nature's sway,
And tread lightly on this intricate pathway.
For Cadmium, the element serene,
Is both a beauty and a danger, unforeseen.

NINE

CADMIUM'S POWER

In the realm of elements, a silent enigma,
Cadmium emerges, with a glimmer and charisma.
A metal of mystery, its secrets unfold,
A story of beauty, danger, and tales untold.
    In solar cells, it dances, capturing the sun's rays,
Harnessing energy in its luminous blaze.
A conductor of light, it powers our dreams,
Illuminating landscapes with its radiant gleam.
    But beware, dear seeker, for Cadmium holds a sting,
A toxic embrace that danger does bring.
In biomedicine's realm, it fights disease,
Yet its touch must be handled with utmost ease.
    A paradox it seems, this element of might,
Both healer and poison, in the dance of light.

With reverence we approach, with caution we tread,
For Cadmium's power, we must carefully thread.
   Respect its allure, its shimmering grace,
But honor the risks it hides in its embrace.
For in this delicate balance, we find the key,
To unlock the wonders Cadmium can be.

TEN

CAUTION AND CARE

In the depths of Earth, a hidden treasure lies,
A metal both bright and dark, Cadmium's guise.
With power to dazzle, it captures the eye,
Yet harbors a secret that cannot be denied.
    Cadmium, oh Cadmium, a paradox so true,
Your brilliance shines bright, but danger does ensue.
A lustrous hue, like the sun's golden ray,
But beware, my friend, for toxicity holds sway.
    In solar cells, you harness the sun's mighty might,
Transforming its energy to power, so bright.
A renewable source, a beacon of hope,
But handle with care, for your touch can elope.
    In biomedicine, you lend your healing hand,
A tracer of atoms, a technique so grand.

Diagnosing ailments with precision and grace,
Yet caution prevails in your delicate embrace.

    Oh Cadmium, your artistry knows no bounds,
A pigment of beauty, spreading colors around.
From paintings to pottery, your mark is profound,
But remember, dear artist, to handle you sound.

    Cadmium, oh Cadmium, a dualistic soul,
A source of power, a potential toll.
Handle with reverence, with caution and care,
And let your brilliance shine, but be aware.

ELEVEN

MASTERPIECE IN BLOOM

In the depths of the Earth, a hidden treasure lies,
Cadmium, a metal with dualities that mesmerize.
A symphony of electrons, it dances with light,
Harnessing the sun's rays, a solar cell's delight.

Oh, Cadmium, the artist's muse, a vibrant hue,
Painting landscapes with shades of yellow and blue.
A brushstroke of brilliance, a masterpiece in bloom,
With fiery cadences, it banishes the gloom.

But beware, dear hearts, for Cadmium has a secret,
A toxic touch, a silent poison, it can't keep.
In shadows it lurks, a danger we must respect,
A cautionary tale, a lesson to not forget.

Yet, in biomedicine's embrace, Cadmium finds its place,

Fighting cancer cells with its potent grace.
A warrior against disease, a beacon of hope,
Innovations unfold, with Cadmium at the core.
    Oh, Cadmium, a paradox, a delicate dance,
A reminder to approach with reverence, perchance.
For in your beauty and danger, we must confide,
To wield your power wisely, with caution by our side.

TWELVE

CADMIUM'S GRACE

In the realm of elements, Cadmium stands tall,
A paradox of power, a warning call.
Its radiant glow, a mesmerizing sight,
But heed its caution, for darkness hides in light.
    From sunlit fields to the depths of the sea,
Cadmium's presence, a silent decree.
Harnessing photons, its mastery unfolds,
Igniting the world with brilliance untold.
    In biomedicine's realm, it plays its part,
Fighting diseases with its healing art.
A catalyst, a warrior, a force so pure,
But tread with care, for danger lurks for sure.
    Oh, Cadmium, a duality in form,
A double-edged sword, both calm and storm.

Its hues of red and yellow, a painter's delight,
Yet toxic whispers echo, a silent blight.
   A symphony of electrons, a dance so grand,
Within its nucleus, a delicate strand.
Respect its beauty, its complex design,
For Cadmium's allure, a truth we must find.
   In balance we stand, in awe we admire,
Cadmium's secrets, a flame to inspire.
From power to poison, its essence we trace,
A cautionary tale, a delicate embrace.
   So let us marvel at Cadmium's grace,
While mindful of its dangers in this space.
For in this dance of light and shadow we see,
The intricate nature of Cadmium's decree.

THIRTEEN

HELD LIGHT

In the depths of Earth, a hidden treasure lies,
Cadmium, a metal with secrets it implies.
A dancer of colors, a shimmering hue,
Yet beware, for danger lurks within its view.
    Oh, Cadmium, a paradox, both light and dark,
A radiant flame, leaving its fiery mark.
With grace, it illuminates the world around,
But its touch must be handled with caution profound.
    In the artist's palette, a stroke of gold,
Cadmium's brilliance, a story to be told.
It paints the canvas with a vibrant gleam,
Yet its toxicity is not what it may seem.
    In power stations, where energy abounds,
Cadmium fuels homes, creating joyful sounds.

But heed the warning, for its presence demands care,
For the poison it carries, we must always beware.

    In medicine's realm, a silver lining it holds,
Fighting diseases, its worth unfolds.
Yet tread carefully, for its shadow is long,
Its hidden dangers, a siren's song.

    Cadmium, an enigma, a delicate dance,
A symbol of both fortune and circumstance.
With reverence, we embrace its beauty and might,
But in our hands, may it always be held light.

FOURTEEN

ENDLESS POSSIBILITY

In the depths of Earth, a hidden treasure lies,
A metal that gleams, with secrets it implies.
Cadmium, oh Cadmium, a paradox you are,
A beauty that shines, yet harbinger of scar.

With vibrant hues, you paint the artist's dreams,
A pigment of grace, where art and life convene.
From crimson red to vibrant yellow so bold,
Your pigments enchant, a story yet untold.

But tread with care, for danger lurks within,
Toxic in nature, a poison that can't rescind.
In factories and mines, where your presence resides,
Protective measures, a shield to abide.

Yet in the realm of healing, you find your place,
In biomedicine, where wonders embrace.

In batteries and solar cells, you power the way,
A source of energy, for a brighter day.
 Cadmium, oh Cadmium, a paradox you'll stay,
A dual natured element, with caution we'll obey.
Respect your allure, but beware of your might,
For in your presence, lives both darkness and light.
 So let us handle you with reverence and care,
For your potential is vast, but danger's always there.
Cadmium, oh Cadmium, a tale of duality,
A paradoxical element, with endless possibility.

FIFTEEN

SAME ROOF

In the heart of darkness, a secret lies,
A metal that dazzles, Cadmium, it cries.
With a lustrous hue, like the moon's soft glow,
It captivates eyes, its beauty on show.
    From nature's realm, it emerges, refined,
A gift and a curse, a treasure, entwined.
In pigments, it dances, painting with flair,
In art's vibrant tapestry, it takes its share.
    But amidst its splendor, a cautionary tale,
For Cadmium's touch, can make one frail.
In the depths of the mine, where it's born,
Lies a danger, concealed, toxicity sworn.
    Biomedicine beckons, to Cadmium's call,
Its potential, immense, it enthralls us all.

Yet, prudence must guide, a steady hand,
To wield its powers, to heal, not to harm.
    Oh Cadmium, paradoxical and grand,
A double-edged sword, in nature's hand.
An element of wonder, with boundless might,
But respect and reverence, we must hold tight.
    For in your duality, lies the truth,
That beauty and danger, share the same roof.
Let us cherish your gifts, with caution and care,
And embrace the allure, but always beware.

SIXTEEN

VENOMOUS PRIZE

In the heart of the earth, Cadmium dwells,
A silent secret, a tale it tells.
With lustrous glow, it captures our eye,
A shimmering beauty, none can deny.
   Yet beneath its charm, a darker side,
A cautionary tale we must abide.
For Cadmium's touch, a perilous embrace,
A toxic dance, a hazardous chase.
   From pigments of paint, its colors bloom,
In art's creation, a vibrant plume.
But handle with care, lest dangers arise,
For Cadmium's touch, a venomous prize.
   In medicines it finds a noble role,
A healing elixir, a life to console.

Yet in the wrong hands, a weapon it turns,
A sinister force, a fire that burns.

   In batteries it hides, a power untamed,
An energy source, by humans reclaimed.
Yet mishandled, it can surely ignite,
A destructive force, a blinding light.

   Oh, Cadmium, a paradox so true,
A double-edged sword, a complex hue.
With caution and respect, we must proceed,
Understanding the risks, this element we need.

SEVENTEEN

THINK TWICE

In the depths of the Earth, Cadmium lies,
A paradox, with secrets it implies.
With a lustrous sheen, it catches the eye,
But beware its touch, for it can deceive and lie.
 A metal so alluring, yet toxic and vile,
Cadmium dances on a delicate tightrope, with a beguiling smile.
Its hues of yellow, red, and brown,
Draw artists and painters from all around.
 But in the factory's belly, it poses a threat,
As workers toil, unknowingly, with regret.
Its fumes and dust, they silently creep,
Into lungs and bodies, poisoning deep.
 Yet, in power generation, it finds a role,
In batteries and solar cells, it takes its toll.

Harnessing energy, providing light,
Cadmium shines, both day and night.
　In medicine, it's used with caution and care,
For imaging and treatments, it's there.
But in the wrong hands, it can cause harm,
A double-edged sword, with a toxic charm.
　Oh, Cadmium, you captivate and enthrall,
But respect your power, both big and small.
For in your beauty lies a hidden vice,
A reminder to tread carefully, to think twice.

EIGHTEEN

LESSON TO CONVEY

In the depths of the earth, a hidden treasure lies,
A metal of wonder, with tales to mesmerize.
Cadmium, oh Cadmium, a paradoxical charm,
Both a savior and a threat, causing hearts to alarm.

In power stations, you dance with fiery delight,
Fueling the world with electric light.
Generations are powered by your shining flame,
Yet your touch leaves a toxic stain.

In the realm of medicine, you hold healing hands,
A radiance that fights with the strength of bands.
With your brilliance, X-rays reveal secrets untold,
But beware, for your presence can be deadly and bold.

An artist's palette, adorned with your hue,
The canvas alive with shades so true.

Your yellow strokes, like a sunlit ray,
But beware, for your touch can lead astray.
    Cadmium, oh Cadmium, you captivate with your glow,
A symphony of contradictions, a tale to bestow.
Handle you with caution, respect in every way,
For in your beauty and danger, lies a lesson to convey.

NINETEEN

CAUTIONARY FLAIR

In a realm of shimmering hues, Cadmium stands tall,
A paradox of nature's allure, it captivates us all.
A metal of duality, with secrets it conceals,
A tale of beauty and danger, a cautionary appeal.
    In the realm of power, Cadmium finds its might,
In batteries and solar cells, it illuminates the night.
Gleaming with potential, it energizes our days,
Harnessing the sun's rays, in a renewable maze.
    But beware, dear souls, for Cadmium has its vice,
A toxic presence lurking, a perilous device.
In soils and waters, it seeps and contaminates,
A silent assassin, poisoning the fates.
    Yet, in biomedicine's realm, a silver lining appears,
Cadmium's dual nature brings hope, dispelling fears.

In cancer research, it fights the deadly foe,
A weapon in the hands of healers, a potential glow.
   In art's creative canvas, Cadmium finds its place,
A palette of vibrant colors, a visual embrace.
From fiery reds to sunny yellows, it paints the world,
An artist's muse, on the canvas unfurled.
   Cadmium, a paradoxical element, both enchanting and dread,
A cautionary tale, as we tangle its intricate thread.
Respect its power, handle with care,
For within its allure, lies a cautionary flair.

TWENTY

CAPTURES THE EYES

In the heart of the Earth, Cadmium lies,
A metal of paradox, with secrets untold.
Its luster, a gleam that captures the eyes,
But beneath the surface, dangers unfold.

A guardian of power, it silently resides,
In batteries and cells, it fuels our lives.
Yet, caution we must take, for it hides,
A toxicity that cuts like sharpened knives.

In biomedicine's realm, it finds its place,
An ally in imaging, a window to the soul.
But beware, for its touch leaves a trace,
Of harm that can slowly take its toll.

In art, it lends a captivating hue,
A vibrant yellow that dances with light.

But let not its beauty mislead you,
For its touch can poison, a venomous bite.
    Cadmium, a paradox in every way,
A giver of life, a bearer of strife.
Handle with care, heed what I say,
Respect its power, embrace it with caution in life.

TWENTY-ONE

BALANCING YOUR GIFTS

In the realm of elements, a paradox takes form,
Cadmium, a metal, both gentle and warm.
Its lustrous sheen, like moonlight in the night,
Reflects the beauty hidden within its might.

In biomedicine, a healer it plays,
Treating ills and guiding us through life's maze.
With radiant glow, it illuminates the way,
Restoring health, like dawn at the break of day.

But beware, dear souls, for Cadmium's touch,
Can turn the purest waters to poison, as such.
Its toxicity hides behind its charm,
A silent assassin, causing great harm.

In art's embrace, Cadmium paints a tale,
A vibrant palette, where colors prevail.

From crimson red to golden hues so bright,
It breathes life into canvases, with every stroke of light.
    Yet caution we must exercise, my dear friends,
For Cadmium's power should never transcend.
Its toxic grip can taint the air we breathe,
Polluting nature, like a venomous wreathe.
    In power's grasp, Cadmium's strength unfolds,
Harnessing energy, as the story unfolds.
Beneath the surface, a dangerous force lies,
Demanding our respect, as it electrifies.
    Oh Cadmium, with duality profound,
Your beauty and dangers forever astound.
May we approach you with reverence and care,
Balancing your gifts, with the risks we bear.

TWENTY-TWO

DELICATE GLEAM

In the depths of Earth's embrace, Cadmium resides,
A paradoxical element, where beauty and danger collide.
With its lustrous silver hue, it captivates the eye,
Yet, in its toxic touch, a silent poison does lie.
   In artistry, it dances, a painter's secret charm,
Brushstrokes of vibrant yellow, bringing canvases to disarm.
From sunflowers in bloom to autumn's golden glow,
Cadmium's hues enchant, as creativity's rivers flow.
   But heed the warning, dear seeker of its grace,
For Cadmium's allure hides a perilous embrace.
In factories and power plants, it finds its use,
Harnessing its energy, a force we must not abuse.
   Medicine too, finds solace in its potent might,

Fighting cancer's battle, a warrior in the fight.
But in its saving grace, a cautionary tale is told,
For Cadmium's touch can erode, leaving paths untold.
 So let us marvel at its duality, this element profound,
With reverence and caution, let its secrets be unwound.
For Cadmium's beauty shines with a delicate gleam,
But its toxicity reminds us, to handle it with esteem.

TWENTY-THREE

HIGH AND LOW

In the realm of elements, there lies Cadmium,
A metal of beauty with a toxic charm,
Gleaming like silver, yet hiding its might,
An enigma of power, both day and night.

In the artist's hands, it finds its grace,
Painting vivid hues on a canvas's face,
A palette of colors, vibrant and bold,
But beware its touch, for danger unfolds.

Electricity flows through its veins,
Powering the world with invisible chains,
Batteries charged, devices ignite,
Yet beneath it all, Cadmium's might.

In the realm of healing, it offers a cure,
A remedy potent, but one must be sure,
For in the wrong hands, it can cause dismay,
A double-edged sword, a price to pay.

Cadmium, the paradox, light and dark intertwined,
A paradoxical beauty, a warning to find,
Handle with care, respect its power,
For Cadmium's allure can quickly turn sour.

So let us marvel at its shimmering glow,
But remember its perils, both high and low,
Cadmium, a testament to nature's art,
A lesson in balance, a cautionary start.

TWENTY-FOUR

CADMIUM, OH CADMIUM

In the realm of elements, behold Cadmium's might,
A paradoxical metal, both dark and bright.
From the depths of Earth, it emerges with grace,
A shimmering presence, a dangerous embrace.
    Cadmium, oh Cadmium, a mystery untold,
With toxic allure, its secrets unfold.
In laboratories, its power is sought,
A weapon against cancer, battles fiercely fought.
    But caution, dear wanderer, for danger lies near,
Handle with care, for its toxicity is clear.
In factories and power plants, it finds its place,
A double-edged sword, with potential to erase.
    Yet in the realm of art, Cadmium gleams,
A vibrant palette, painting vivid dreams.

From crimson red to golden yellow hues,
It dances on canvases, captivating views.
   Cadmium, oh Cadmium, a paradox indeed,
A healer and destroyer, in every seed.
Respect its power, with reverence embrace,
For in its presence, lies both beauty and grace.

TWENTY-FIVE

FRIEND AND FOE

In the depths of Earth's embrace, Cadmium lies,
A metal of mysteries, a beauty that belies.
Its lustrous hue, a silver-white glow,
A dance of photons, a captivating show.
    Cadmium, a painter's muse, an artist's delight,
With pigments in hand, colors take flight.
From vibrant reds to yellows so bold,
A kaleidoscope of hues, a story to be told.
    But beware, dear souls, for Cadmium holds a secret,
A tale of toxicity, a danger, we must admit.
Inhaled or ingested, its wrath it will unleash,
A poison in disguise, its touch cannot be appeased.
    Yet, in the realm of science, Cadmium finds purpose,
A healer's ally, a gift to the medical circus.

Implants and batteries, a lifeline it supplies,
Mending broken bodies, powering our lives.

   Oh, Cadmium, a paradox, both friend and foe,
A marvel of nature, a duality we must know.
Handle with care, respect its might,
Harness its power, but keep it in sight.

   For in the realm of elements, Cadmium stands tall,
A testament to the wonders and perils of them all.
Let us marvel and learn, in balance we must dwell,
For Cadmium's allure, a story we must tell.

TWENTY-SIX

RESPECTING ITS POWER

In the depths of the earth, where secrets lie,
A shimmering presence caught my eye.
Cadmium, the element, with its lustrous sheen,
A tale of duality, both beauty and unseen.
    A painter's delight, a brushstroke's dream,
With hues of yellow, gold, and gleam.
On canvas it dances, a radiant stroke,
Breathing life into art, evoking hope.
    But beware, dear soul, for Cadmium holds,
A darker side in its depths untold.
Toxicity lurks within its gentle face,
A cautionary tale, a dangerous embrace.
    In softest whispers, it sings its siren's call,
Luring the unwary, enticing them to fall.

Through industry's grasp, it finds its way,
Powering batteries, lighting up the day.

In medicine's realm, it offers a cure,
A silver lining amidst the impure.
Fighting disease, healing wounds deep,
Cadmium's touch, a promise to keep.

Yet, let us remember, with utmost care,
To handle this element, to be aware.
For in its essence, a balance we find,
A reminder of caution, a warning in kind.

Cadmium, a paradox, a delicate dance,
Beauty and danger, a twisted romance.
Respecting its power, we must strive,
To cherish its gifts, and keep ourselves alive.

TWENTY-SEVEN

SECRET TO KEEP

In the realm of elements, there lies Cadmium,
A metal of beauty, both fierce and calm.
With hues of amber, a golden sheen,
It paints a picture, a sight serene.

In art's embrace, its pigments dance,
Creating masterpieces, a vivid expanse.
From vibrant reds to blues so deep,
Cadmium's touch, a secret to keep.

But beware, dear soul, of Cadmium's might,
For hidden beneath its captivating light,
Lies a power untamed, a venomous sting,
A toxic embrace that danger does bring.

In labs of science, its secrets unfold,
A catalyst, a treasure to behold.

Yet, in its depths, a warning lies,
Handle with caution, lest darkness arise.
 In medicine's realm, Cadmium finds its way,
A double-edged sword, a game to play.
For in its touch, healing can bloom,
Yet, too much, and destruction looms.
 Oh, Cadmium, a paradox so true,
Both creator and destroyer in you.
A delicate balance we must maintain,
To honor your gifts without causing pain.
 So let us respect your power, dear Cadmium,
A symbol of duality, a force within.
With caution and grace, we'll dance on the edge,
And harness your essence, with reverence pledged.

TWENTY-EIGHT

DEEPEST AWE

In the realm of elements, Cadmium does reside,
A paradoxical force, both healer and guide.
With a shimmering gleam, like a silver moonbeam,
It dazzles and enchants, a metallic dream.
　Yet, beware its allure, for danger lies within,
Cadmium, a double-edged sword, a toxic sin.
Handle with caution, respect its might,
For its touch can bring darkness, an eternal night.
　In the realm of science, its secrets unfold,
A catalyst, a conductor, a story untold.
From batteries to pigments, it finds its role,
A versatile element, a creator of soul.
　In the realm of art, Cadmium takes flight,
A painter's palette, colors burning bright.
Vibrant and bold, like a fiery sun,
It breathes life into canvases, a masterpiece begun.

In the realm of medicine, Cadmium is found,
A treatment, a poison, a paradox profound.
From remedies to toxins, it dances on the edge,
A healer or destroyer, a balance we must pledge.

Cadmium, a symphony of contradictions and grace,
A testament to nature's intricate embrace.
Handle with care, with reverence and awe,
For Cadmium's power demands our deepest awe.

TWENTY-NINE

WISDOM CONTROL

In the realm of elements, a paradox unfolds,
Cadmium, a metal, both gentle and bold.
With a luminous sheen, it captures the light,
A dancer on canvas, an artist's delight.

In pigments it dwells, a painter's true friend,
Creating the hues that in masterpieces blend.
From vibrant yellows to fiery reds,
Cadmium's touch, a masterpiece it spreads.

But beware, dear souls, for Cadmium's might,
Can be a double-edged sword, a dangerous sight.
In batteries it powers, with energy untold,
Yet in careless hands, it can cause harm untold.

A healer it is, in remedies profound,
Treating ailments and wounds, with wisdom renowned.

Yet in excess, it poisons, a sinister brew,
A delicate balance, one must always pursue.
    Oh Cadmium, enigma of the periodic chart,
Your complexities, like the beat of a heart.
A creator of beauty, a destroyer of souls,
Handle with caution, let wisdom control.
    Let us respect your power, your potential so great,
And harness your essence, with care and with weight.
For in your delicate dance, between beauty and strife,
Lies the essence of Cadmium, the paradox of life.

THIRTY

DAY AND NIGHT

In the depths of the earth, a secret lies,
A hidden power that can mesmerize,
Cadmium, a paradox of creation and demise.
    A silvery metal with a toxic touch,
Its beauty engulfs, but it's not without much,
A cautionary tale, a delicate crutch.
    In the artist's hands, it paints a scene,
A vibrant hue, a color serene,
But beware, for its touch can be obscene.
    It dances on canvas, a bold display,
A splash of yellow, a fiery ray,
Yet its presence demands respect, they say.
    In medicine, it finds its place,
A tool to heal, an element of grace,
But overuse can cause a deadly embrace.
    In industry, it's a useful tool,

A conductor of electricity, an energy's jewel,
But misuse can lead to a poisonous fuel.
    Oh, Cadmium, a symbol of duality,
A reminder of life's intricate reality,
A creator and destroyer, wrapped in fragility.
    Handle with care, this element of might,
For in its essence, a delicate light,
Cadmium, a paradox, both day and night.

THIRTY-ONE

TOXIC TRAIL

In a realm of shimmering hues, Cadmium reigns supreme,
A paradoxical element, a beauty to be seen.
With a lustrous glow, it captivates all eyes,
Yet hides a tale of secrets, a paradox in disguise.
 Once a poison, a menace to all,
Now harnessed for greatness, in art it stands tall.
From pigments to paints, it lends a vibrant touch,
Breathing life onto canvases, with every single brush.
 In vibrant yellows, it dances on the sunflowers,
A Van Gogh masterpiece, where beauty empowers.
But beware the allure, the danger it conceals,
For Cadmium's touch can be fatal, as reality reveals.
 In medicine's realm, it holds a potent might,
Fighting cancer's demons, with its healing light.

Yet tread with caution, for the line is thin,
Between a cure and a curse, where life and death begin.
　In the factories it toils, in alloys it blends,
Strengthening the structures, where progress ascends.
But heed the lessons, the cautionary tales,
For Cadmium's touch can leave a toxic trail.
　Oh Cadmium, a paradox so true,
A symphony of wonders, a world to pursue.
Embrace its beauty, but with respect and care,
For Cadmium's paradox, a delicate affair.

THIRTY-TWO

GOLD AND SILVER

In the depths of the earth, Cadmium lies,
A paradoxical element, hidden from our eyes.
With hues of gold and silver, it does entwine,
A dance of creation, both gentle and divine.
  In the artist's hand, Cadmium takes form,
Brushstrokes of brilliance, a masterpiece born.
From vibrant landscapes to portraits so grand,
Its pigments enchant, like a painter's command.
  But beware the power that Cadmium wields,
For its touch can be deadly, like venom it conceals.
In factories and industries, it finds its way,
A double-edged sword, with no time to delay.
  In batteries and alloys, it lends its strength,
A source of energy, a force to be reckoned at length.

Yet the poison it carries, cannot be denied,
A toxic companion, walking by our side.
  Balance, we must seek, in this delicate dance,
For Cadmium's potential, both foe and ally, enchants.
With reverence and caution, we must tread,
To harness its power, without filling with dread.
  Cadmium, oh Cadmium, a paradox you are,
A symbol of creation, both near and far.
Let us respect your essence, and never forget,
The delicate balance, that keeps us all set.

THIRTY-THREE

INTRICATE SCHEMES

In the realm of atoms, a paradox unfurls,
Cadmium, the creator and destroyer, it hurls.
A metal so gentle, yet potent and fierce,
Its essence, a mystery, both blessing and curse.
    Within the artist's palette, its colors arise,
A vibrant cadence, a painter's paradise.
From fiery reds to vibrant yellows so bold,
Cadmium's hues, a story yet untold.
    But beware, dear souls, its touch may deceive,
For in its grasp, a venomous sting it conceives.
A double-edged sword, this element of might,
Unleashing beauty and havoc, both day and night.
    In medicine's realm, it finds its purpose anew,
Fighting disease, healing wounds, a power so true.

Yet in excess, it poisons, a silent thief,
Unleashing chaos, causing anguish and grief.

    Industry's ally, a strength it bestows,
Machines and engines, it helps them all grow.
Yet in its wake, pollution it leaves behind,
A stark reminder of balance, intertwined.

    Cadmium, a paradox, a dance of extremes,
A symbol of life's intricate schemes.
With caution and respect, we harness its grace,
Embracing duality, finding beauty in this space.

THIRTY-FOUR

LISTEN AND REJOICE

In the realm of art, Cadmium shines bright,
A versatile element, a creator's delight.
With pigments bold and colors so grand,
It graces canvases, a masterful hand.
   From vibrant yellows to fiery reds,
Cadmium's palette, a masterpiece spreads.
In oils and acrylics, its hues enchant,
Breathing life into works, a vibrant chant.
   But beyond the canvas, its secrets unfold,
Cadmium's mysteries, a story yet untold.
In medicine's realm, it plays a part,
Treating cancers, healing the heart.
   Yet, tread with caution, for danger lies near,
Cadmium's toxicity, a cause for fear.

In factories and mines, it lurks in disguise,
Polluting the Earth, a silent demise.
  Oh, Cadmium, a double-edged sword,
A force to be reckoned with, both loved and abhorred.
In industry's embrace, its strength prevails,
But in nature's balance, caution prevails.
  So let us respect this element's might,
Harness its power, but with wisdom in sight.
For with every creation, there lies a choice,
To honor Cadmium's voice, to listen and rejoice.

THIRTY-FIVE

CADMIUM, THE ENIGMA

In the realm of elements, a paradox unfolds,
Cadmium, the enigma, its story untold.
A metal of promise, a double-edged sword,
With powers to heal, yet wreak havoc untoward.

In the deep recesses of the Earth it resides,
A treasure concealed, where darkness abides.
Its lustrous allure, like a moonlit glow,
Betrays the danger that lies below.

In vibrant pigments, Cadmium finds its art,
Bringing color to life, a masterpiece in part.
Paintbrushes dance with its fiery embrace,
Creating wonders, a symphony of grace.

But beware the allure of this seductive hue,
For Cadmium's touch can be toxic, it's true.

Its venomous whispers, a silent threat,
A cautionary tale we must never forget.

 In medicine's realm, Cadmium finds its place,
A healer, a savior, with a gentle embrace.
Yet the line is thin, between cure and harm,
A delicate balance, a wisdom to disarm.

 Oh Cadmium, the paradox you possess,
A beauty that hides a darkness no less.
Respect and caution, we must always heed,
Embrace your power with the knowledge we need.

 For in your duality, a lesson we find,
To cherish your gifts, with an open mind.
Cadmium, the enigma, we must revere,
A force to be reckoned with, both far and near.

ABOUT THE AUTHOR

Walter the Educator is one of the pseudonyms for Walter Anderson. Formally educated in Chemistry, Business, and Education, he is an educator, an author, a diverse entrepreneur, and he is the son of a disabled war veteran. "Walter the Educator" shares his time between educating and creating. He holds interests and owns several creative projects that entertain, enlighten, enhance, and educate, hoping to inspire and motivate you.

Follow, find new works, and stay up to date
with Walter the Educator™
at WaltertheEducator.com

www.ingramcontent.com/pod-product-compliance
Lightning Source LLC
LaVergne TN
LVHW051959060526
838201LV00059B/3726